HERB AND SPICE SEEDS
A Pictorial Field Guide

Terry A. Woodger

Universal-Publishers
Boca Raton

Herb and Spice Seeds:
A Pictorial Field Guide

Universal-Publishers
Boca Raton, Florida
USA • 2011

ISBN-10: 1-61233-043-6
ISBN-13: 978-1-61233-043-3

www.universal-publishers.com

Library of Congress Cataloging-in-Publication Data

Woodger, Terry A.
  Herb and spice seeds : a pictorial field guide / Terry A. Woodger.
    p. cm.
  Includes bibliographical references and index.
  ISBN-13: 978-1-61233-043-3 (pbk. : alk. paper)
  ISBN-10: 1-61233-043-6 (pbk. : alk. paper)
  1. Herbs--Seeds--Harvesting. 2. Spice plants--Seeds--Harvesting. 3.
Seeds--Cleaning. 4. Herbs--Seeds--Storage. 5. Spice plants--Seeds--
Storage. 6. Herbs--Seeds--Pictorial works. 7. Spice plants--Seeds--
Pictorial works. I. Title.
  SB118.3.W67 2011
  631.5'21--dc23

                                                    2011032346

ACKNOWLEDGMENTS

A book of this nature can not be written without the assistance of family and friends. I would like to acknowledge the following people who assisted in many ways to help make this book a reality.

First, I'd like to thank my wife and children for their support and encouragement, without which this book would never have been completed.

I'd also like to thank the following people for their assistance:

Andrew Leighton
Clive and Cathy Grimshaw
Graham and Kathy Evans
Jacqueline Weight
Members of the Townsville Herb Society 2010
Rick and Helen Dyer
Ron and Lynda Roundhill

Disclaimer:
Plants have many ways in which they protect themselves from damaging organisms. This protection is found in thorns, sap, toxins, etc. Although the collection, cleaning, and storage of seed can be a rewarding experience, the author takes no responsibility for injury or illness that results from these activities.

CONTENTS

INTRODUCTION

Seeds are an exciting and beautiful component of gardening herbs and spices. Productive herb gardens that provide a great sense of fulfillment can be created from just a handful of these treasures. Although some of these plants can propagate by other means, such as bulbs or division, seeds are the principal way in which herbs and spices reproduce.

This book covers the basics involved in the collection, cleaning and storage of seeds. Although bulbs, corms, and other plant parts can be collected and stored, they are not covered here, so as to not detract from the focus of this volume.

As the number of plants grown for herbs and spices is truly staggering, it is impossible to cover them all. In this field guide, we discuss the most common plant families, including examples of the types of seeds that may be encountered. Where possible, several genera within each family are discussed.

This book uses a system whereby plant family names are all written in capitals (BORAGINACEAE), the common names that are not written within the text are in bold (**Borage**), and the botanical names are written in italics (*Borago officinalis*).

In botany, it is the characteristics of the flowers that determine the genera and family to which a plant belongs. This can become extremely complicated, so this field guide makes no mention of the flower types or their individual differences.

Also discussed are a number of methods that can be utilized in the collection of seeds. No one method can be used to collect them all, so different techniques have been developed over time to successfully gather all of the species that are encountered, both in the home garden as well as in the field.

The same development of techniques applies to the cleaning of seeds. There are a number of ways in which common household items can be used effectively to clean seeds. Several of these items are explained in Chapter 3: How to Thresh and Clean Seed.

Storing seeds for use next season can be fraught with hidden problems, such as molds and seed-borers. Chapter 4: The Storage of Seed explains some appropriate methods and procedures that should be followed to avoid disappointment and loss of seed.

Most seeds collected from the garden are suitable for storage from one year to the next, and many of these can be successfully stored at home for many years.

Some things to consider when collecting seeds are the quantity and the number of plants from which they are collected.

Collecting seed from only one fruit on one plant over several seasons can have unforeseen consequences, such as the collection of seeds from small fruit, especially from only one plant. This can produce smaller plants with smaller fruit. This is often referred to as "linebreeding" and can, over time, lead to complete genetic breakdown and a loss of viability.

The collection of seed is not always a simple matter, and it is very important that, when possible, you gather seed from several plants to maintain their genetic stability.

Size does matter, and in the case of herb and spice seeds, it is always best to collect seed from the biggest fruit or healthiest plants.

The exception would be if you are after a specific genetic trait. For example, seed selection would vary for a plant desired as a bush if that plant is usually a climbing variety.

Now let's look at the question of, what is a fruit?

When considering plants, a fruit is any structure that produces seed, but not always something that is edible. The following pages outline the types of fruit that will be encountered in the collection of seeds from herbs and spices.

One of the first things that should be learned in the collection of seed is the differences between the various fruits that you will encounter, as this will determine how to collect and clean the seed.

The methods used for the collection of seeds do not require an intimate knowledge of fruits. Knowing the basics as outlined here will aid in the collection, cleaning, and storage of viable seed.

Achene: An achene is a small, single-seeded, dry fruit usually formed in clusters of the ASTERACEAE (daisy) family. Fruit of this group include the marigold, and the thistle.

Milk Thistle
Sonchus oleraceus

Berry: A berry is a fruit with seeds contained within a fleshy or dry pulp. Berries contain two or more seeds, and include coffee and black elders.

Coffee
Coffea
arabica

Pomegranate
Punica
granatum

Capsule: A capsule is a dry seed case that opens along several seams and can be papery thin to hard and woody. Some open at maturity, releasing their seeds, while others remain closed. Capsules can contain anything from one to over one hundred seeds and are found in many plant families including the orchids, poppies, and onions.

Chives
Allium schoenoprasum

Drupe: A drupe is a fleshy or dry fruit containing one seed, often referred to as a "stone," hence the origin of the term "stone fruits." These include the neem, nutmeg, olive, and pepper.

Neem *Melia indica*

Drupelet: A drupelet is a small drupe, such as an individual segment of a blackberry or raspberry.

Etaerio and Sorosis: These are aggregates or fruits from either a single flower or a group of flowers. Etaerio fruits include the raspberry, blackberry, and strawberry. Sorosis fruits include the mulberry and pineapple. For the purposes of this book, the differences between these fruits are of little importance.

Strawberry *Fragraria vesca*

Grain: "Grain" is a term used to describe the husk-covered seeds from the family POACEAE (grasses, cereal crops). Several of these plants are used as herbs and spices such as lemon and citronella grass.

Grain head of cereal crop

Hesperidium: A hesperidium is a berry with a thick rind and is made up of segments. These are fruits of the citrus family and include the orange, lime, and lemon, and many others.

Lemon *Citrus limon*

Hip: A hip is a term used to describe the fruit of a rose. It is essentially a dry berry.

5

Rose Hip *Rosa* sp.

Legume: Legumes split into two equal halves upon drying. Legumes include all the beans and peas. They can be thin and papery, or thick and woody. The tamarind and carob are examples of legumes.

Tamarind *Tamarindus indica*

Pepo: The pepo is a berry with a hard rind. This includes all of the cucurbits, such as the gourds and luffa.

Luffa *Luffa acutangula*

Schizocarp: These fruits are from the flannel flower family (APIACEAE). All are small dry fruits that are made up of two capsules attached at the base, these split into two equal parts at maturity.

Dill *Anethum graveolens*

Silique: These fruits are all from the mustard family (BRASSICACEAE). All have dry, woody pods with a partition down the center. The pods open at maturity and release their seeds as they begin to dry.

Mustard *Brassica juncea*

CHAPTER 1
THE COLLECTION OF SEED

Seed should be collected when the weather is fine and the plants are dry.

If you have been able to collect seed pods or capsules when they are dry, then it is only a matter of cleaning. However, if the collected material is wet or damp, then mold may become an issue. Wet collected material can be spread out on a tarp in the sun to dry. If this is not possible, spread them near some sort of warmth, such as a heater, and allow them to fully dry prior to cleaning.

Removing seeds from fleshy berries and drupes is usually a simple matter if they are fully mature.

When transporting seed overseas or across borders, a declaration of the species and amounts is often required, as some species are prohibited. If the seed is not thoroughly cleaned it may be confiscated.

When seed is required for personal use and cleanliness is not so important, keep in mind that insects and molds can still become a problem, so it remains important to clean all seed as thoroughly as possible.

There are a number of ways to collect seed; it is a matter of selecting the one which works best for you and the plant species from which you are collecting. The use of a bucket or two when collecting fruits, capsules and flower heads is essential. Collecting large amounts of seed by hand or only small amounts from a few plants can become cumbersome without something to put it in; you can use bags, but they can often become more annoying than helpful unless you put the bags in the bucket.

Many herbs and spices produce flower stalks that make collection easy. You can gather them in two ways depending on which one is most convenient to the requirements of the individual. You can harvest the individual seed capsules as they ripen or you can place a material bag over the entire seed head and tie the mouth of the bag off. The seed head can be cut off immediately or left for a while on the plant. The risk of leaving seed heads on the plant is moisture from rain or irrigation. If the seed becomes wet, mold and spoilage will become a major threat.

Another useful method is to place a ground sheet under the plant and tap or shake the plant so that the seeds fall onto the ground sheet. The severity of the tap or shake will often depend on the maturity of the seed and the species of plant.

Avoid collecting seed or damaged fruit from the ground as they can be contaminated with mold or insects.

By following the methods outlined for each herb and spice variety, the collection of viable seed should be achievable throughout the seasons.

Chapter 2
Herbs and Spices

Herbs are the plants we use for medicines, flavorings, and seasonings to enhance the wellbeing of everyday life. Herbs represent many more families of plants than the vegetables, and are utilized by all civilizations and cultures throughout the world. These plants have been mentioned in literature from ancient times.

Several of the plants used as vegetables are also considered herbs and spices, such as fennel and onions. These plant families are also covered as necessary in this chapter.

The collection of seed from many of the herb families is covered in this chapter, including the various fruiting forms that will be encountered. Collect the best quality fruits and seeds and use the various methods outlined in the introduction for collecting.

Family: ANACARDIACEAE
Common name: **Cashew**
Number of genera: 70
Number of species: 600
Origin: Tropical and subtropical
Plants: Trees and shrubs

The fruits are drupes; they range from small and papery to large and fleshy. The color of the seed ranges from white to grey. The seeds of the pepper trees are examples of spices in this family and are collected whole when fully colored, pink or red.

Peppercorn *Schinus molle*

Family: APIACEAE, UMBELLIFERAE
Common name: **Flannel Flowers**
Number of genera: 300
Number of species: 3000
Origin: Worldwide
Plants: Herbs with some shrubs

The dry fruit usually splits into 2 seeds upon maturity. These seeds are light yellow to brown. These plants produce a flower spike containing a cluster of small white flowers. The seeds can be collected individually or the entire seed head can be cut from the plant and placed upside down in a fabric or paper bag and left to fully dry before threshing.

Dill *Anethum graveolens*

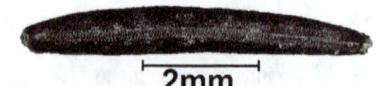

Chervil *Anthriscus cerefolium*

Parsley
Petroselinum
sativum

Anise
Pimpinella
anisum

Cumin
Cuminum
cyminum

Caraway
Carum
carvi

Coriander
Coriandrum
sativum

Lovage
Levisticum
officinale

Angelica
Archangelica
archangelica

Mitsuba
Cryptotaenia
candensis

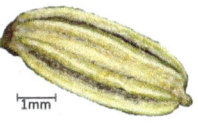

Celeriac
Apium
graveolens
reapceum

Fennel
Foeniculum
vulgare
dulce

Family: ASTERACEAE,
COMPOSITAE
Common name: **Daisy**
Number of genera: 1100
Number of species: 20,000
Origin: Worldwide
Plants: Perennial herbs with a few
trees and shrubs

The daisies are the largest of the plant families. The fruit is a burr, drupe, or a group of closely packed singular seeds (achene).

The seeds can be slightly or largely elongated with grooves, pits, or wings. Some are burrs and many have tufts of long hairs.

Usually, there is a single flower stem producing a cluster of flowers with numerous seeds. The seeds can be collected daily by picking each dried flower head. Alternatively, the entire stem can be collected once most of the flowers have died.

Yarrow
Achillea
millefolium

Wild Chicory
Cichorium
intybus

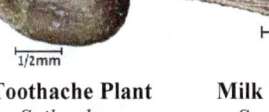

Toothache Plant
Spilanthes
acmella

Milk Thistle
Sonchus
oleraceus

9

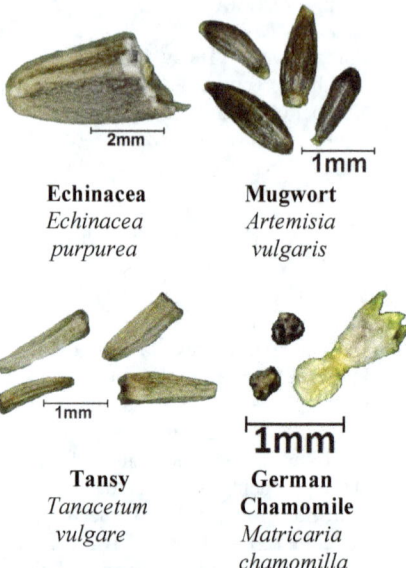

Echinacea
Echinacea purpurea

Mugwort
Artemisia vulgaris

Tansy
Tanacetum vulgare

German Chamomile
Matricaria chamomilla

Family: BORAGINACEAE
Common name: **Borage**
Number of genera: 146
Number of species: 2000
Origin: Worldwide
Plants: Trees, shrubs, and herbs

The fruits are capsules, or sometimes drupes or nutlets. Some seeds have barbed hooks that attach to passing animals or objects. The capsules will often open and release their seed before they are noticed. Collect the capsules as they are encountered and save until a suitable quantity is ready to be sieved of chaff.

Borage *Borago officinalis*

Family: BRASSICACEAE, CRUCIFERAE
Common name: **Mustard**
Number of genera: 375
Number of species: 3200
Origin: Worldwide
Plants: Herbs with some sub-shrubs

This family is formally called CRU-CIFERAE. The fruits called silique are cylindrical capsules. They dry prior to splitting into two equal halves with a partition down the center, releasing numerous seeds. The seeds are brown to black in color. The seed pods can be collected individually or the entire seed head can be collected and placed upside down in a paper or fabric bag.

Cress
Barbarea vulgaris

Shepherd's Purse
Capsella bursa-pastoria

The **Mustards** are grown for their seed, so the whole plant is left until the seed is mature. The seed pods can be collected individually. The plants are often too large to place in a bag, so the whole plant can be cut down and placed on a ground sheet to dry. The pods are then threshed and the chaff is removed by sieving.

White

Brown

Mustard
Brassica juncea

Family: CAESALPINIACEAE
Common name: **Bean**
Number of genera: 150
Number of species: 2200
Origin: Worldwide
Plants: Trees and shrubs with some herbs and climbers

The fruit generally resembles the typical bean shape, and can be woody through to thin papery pods with several to numerous seeds. The seeds can be spherical to flat and oval, as well as cylindrical. They are generally reddish, grey, brown, or black, and sometimes mottled.

In the case of the tamarind, the seed is contained in a sticky pulp within a brittle capsule. The seed is mature when the capsule is dry and brittle.

5mm

Tamarind *Tamarindus indica*

Family: CANNABACEAE
Common name: **Hemp**
Number of genera: 2
Number of species: 4
Origin: Temperate
Plants: Herbs and shrubs

The fruit is a nut or dry capsule. Each contains a single brown seed. The seed is collected once the nut or capsule dries. Thresh lightly to release the seed and clean before storage.

2mm

Cannabis *Cannabis sativa*

Family: CAPPARACEAE
Common name: **Caper**
Number of genera: 45
Number of species: 800
Origin: Tropical and subtropical
Plants: Trees, shrubs, herbs, climbers

The fruit is an oval or cylindrical dry capsule that houses one to numerous seeds, or a fleshy berry containing one to numerous seeds in a sticky pulp. The seeds are round and flattened, or can be smooth to pitted. The fruits are collected whole. The seeds can be removed and cleaned by gently pressing the pulp through a sieve with your fingertip and slow running water.

1mm

Caper *Capparis spinosa*

Family: CARYOPHYLLACEAE
Common name: **Pink**
Number of genera: 75
Number of species: 2000
Origin: Worldwide
Plants: Herbs with some shrubs

11

The fruit is a very small round capsule (about 0.5 – 3.5mm) with one to numerous seeds that are slightly flattened. The seeds are collected as the capsule dries. The capsules require gentle threshing to release the seed before sieving out the chaff.

Soapwort
Saponaria officinalis

Chickweed
Stellera media

Family: CUPRESSACEAE
Common name: **Cedar**
Number of genera: 30
Number of species: 140
Origin: Worldwide
Plants: Trees and shrubs

The fruits are woody cone-like structures containing one to many winged or non-winged seeds. Collect the cones as they start to change color near maturity and allow the cones to open and drop the seeds as they dry.

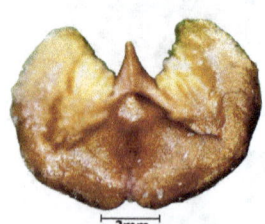

Cypress Pine
Callitris intratropica

Family: EUPHORBIACEAE
Common name: **Spurge**
Number of genera: 300
Number of species: 7500
Origin: Worldwide
Plants: Herbs and shrubs with a white milky sap.

The fruit is usually a dry capsule that splits lengthwise upon drying; it can also be a drupe or berry. The seeds can be round, oval, or triangular and can be smooth or pitted. Collect the seeds as the capsules dry and begin to split; the seeds should have a glossy sheen when collected fresh.

Sweetleaf
Sauropus androgynous

Sweet Cassava
Manihot dulcis

Caster Oil Plant
Ricinus communis

Family: GERANIACEAE
Common name: **Geranium**
Number of genera: 5
Number of species: 750
Origin: Worldwide
Plants: All are herbs

The geraniums are known by their distinctive bird's bill seed pod. The seed can be oval, cylindrical with a corkscrew- like appendage, or cylindrical with fine hairs. Collect the whole pod and allow time to dry fully before threshing and sieving the seed and chaff.

Herb Robert
Geranium robertianum

Family: ILLICIACEAE
Common name: **Star-Anise**
Number of genera: 1
Number of species: 42
Origin: Tropical to subtropical
Plants: Small trees and shrubs

It is the star anise that is of interest in this family. The fruit is a star-shaped capsule with 8 segments, each containing a single seed. The seed can be threshed out of the capsule, or each seed can be removed individually.

Star Anise *Illicium verum*

Family: LAMIACEAE, LABIATAE
Common name: **Mint**
Number of genera: 180
Number of species: 3500
Origin: Worldwide
Plants: Aromatic herbs with some trees

The fruit is usually a dry capsule with four seeds; these seeds are very small and can be collected as the capsules dry. Look at the area where the flowers form and take note of the small capsule that develops. When this dries and turns brown the capsules can be collected. Alternatively, place a ground sheet under the plant and check for seeds daily.

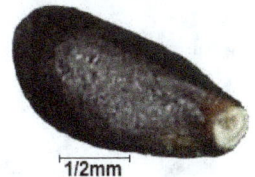

Lemon Balm
Melissa officinalis

Basil   **Java Tea**
Ocimum spp.   *Orthosiphon*
                 aristatus

Motherwort   **Hyssop**
Leonurus   *Hyssopus*
cardiaca   *officinalis*

Rosemary
Rosmarinus officinalis

Lavender
Lavendula vera

Summer Savory
Satureja hortensis

Horehound
Hyptis suaveolens

Nutmeg Bush
Iboza riparis

Catmint
Nepeta cataria

Pineapple Sage
Salvia elegans

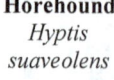

Three-in-one Spice
Solenostemon amboinicus

Wild Thyme
Thymus serpyllum

Patchouli
Pogostemon cablin

Mint *Mentha* spp.

Family: LILIACEAE
Common name: **Lily**
Number of genera: 280
Number of species: 4000
Origin: Worldwide
Plants: Perennial herbs with some evergreen shrubs and climbers

The fruits can be capsules or colorful berries and are harvested once dry or fully ripened. The seeds of this family are generally smooth, with some being winged and others having a variety of ornamentation. The seeds can be shiny orange to red, brown, or black. Collect the berries, and then

Margoram
Origanum majorana

Skulcap
Lateriflora scutellaria

clean away the pulp using a dry cloth. The capsules can then be lightly threshed and the chaff sieved out.

5mm

Barbados Aloe *Aloe vera*

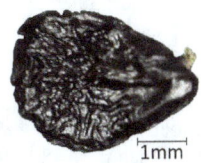

1mm

Chives
Allium schoenoprasum

Family: LINACEAE
Common name: **Flax**
Number of genera: 14
Number of species: 250
Origin: Northern Hemisphere
Plants: Herbs with some large trees or shrubs

The fruits are small dry capsules which divide into 5 equal parts containing 2 flattened seeds each. Collect the capsules once they are dry. Thresh the capsules and sieve the seed from the chaff.

1mm

Linseed
Linum usitatissimum

Family: LYTHRACEAE
Common name: **Loosestrife**
Number of genera: 28
Number of species: 660
Origin: Tropical to temperate
Plants: Trees, shrubs and herbs

The fruit is an elliptical to cylindrical capsule which opens lengthwise as it dries. Seeds are small and numerous. In *Lagerstroemia,* the seed is winged and elongated. Collect the whole capsule and thresh before sieving out the seed.

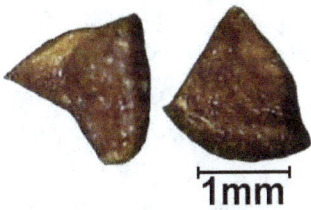

1mm

Henna *Lawsonia inermis*

Family: MALVACEAE
Common name: **Mallow**
Number of genera: 75
Number of species: 1500
Origin: Worldwide
Plants: Herbs and shrubs

The fruits are capsules which split from the top downwards they are sometimes berries. The capsules can be harvested once dry. The seeds are somewhat kidney-shaped, occasionally winged or with horns, and they are brown or black in color.

Collect the capsules using gloves and thresh before sieving out the seed.

The chaff can cause extreme irritation to the skin and may also affect people with breathing difficulties.

Marshmallow
Althea officinalis

Family: MELIACEAE
Common name: **Mahogany**
Number of genera: 51
Number of species: 550
Origin: Tropical and subtropical
Plants: Trees, shrubs with some herbs

The fruit is a capsule, drupe, berry, and sometimes, a nut. The seed is winged or wingless, round or oval. The seed can be collected from the capsules as they open or when the berry or drupe is fully ripe.

Neem *Melia indica*

Family: MORINGACEAE
Common name: **Horseradish-tree**
Number of genera: 1
Number of species: 13
Origin: Tropical to subtropical
Plants: All are trees

The fruit is a woody capsule. Collect the capsules once fully ripe and beginning to split. The seed can be removed from the capsules individually. No threshing is usually required.

Horseradish Tree
Moringa oleifera

Family: MYRISTIACEAE
Common name: **Nutmeg**
Number of genera: 19
Number of species: 300
Origin: Tropical
Plants: Small trees and shrubs

The fruit can be fleshy to non-fleshy drupes or berries. Collect the fruits individually when they are fully ripe. Remove any flesh and pulp before drying and storing.

Nutmeg *Myristica fragrans*

Family: MYRTACEAE
Common name: **Myrtle**
Number of genera: 155
Number of species: 3500
Origin: Tropical and Australasia
Plants: Trees and shrubs

All of these plants have oil glands in their leaves and are predominantly evergreens. Several of these species are grown for their oils. Examples of oils produced are tea tree oil and eucalyptus oil.

The fruits are capsules that can be papery or hard and woody. They split at the front of the capsule upon maturity, releasing their seeds; some capsules may remain unopened for several years before releasing the seeds. The seeds can be saucer-shaped, with a wing all around, be wingless, have a wing on one end, be a pyramid shape, be crescent—shaped, or straight. Color ranges from reddish-brown to brown or black.

Allspice *Pimenta dioica*

Family: OLEACEAE
Common name: **Olive**
Number of genera: 30
Number of species: 600
Origin: Worldwide
Plants: Trees, shrubs, and woody vines

The fruit can be a capsule, drupe, berry, or winged nut. It usually has only one seed. Of interest here is the olive, which is a drupe. Remove the flesh and store the seed once dried.

Olive *Olea europea*

Family: ONAGRACEAE
Common name: **Evening Primrose**
Number of genera: 20
Number of species: 650
Origin: Worldwide
Plants: Trees, shrubs and herbs

The fruits are usually capsules that may or may not open as they dry, but they can also be berries or nuts. The seed quantity ranges from 2 to over 100, depending on which fruit is encountered. Some seeds are non-hairy, others hairy, and some have a tuft at one end. Collect the capsules and thresh to release the seed. Sieve out the chaff and store the seed.

Evening Primrose
Oenothera blennies

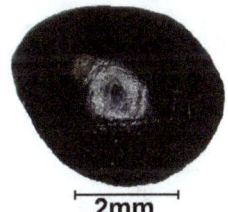

Willow Herb
Epilobium billardierianum

Family: ORCHIDACEAE
Common name: **Orchid**
Number of genera: 730
Number of species: 20,000
Origin: Worldwide
Plants: Epiphytic or terrestrial perennial herbs with some climbers

Of all the orchid species, it is the genera *Vanilla* that is utilized as an herb. There are over 100 species of *Vanilla,* but only 3 are grown commercially; *Vanilla planifolia* is the most preferred.

The fruit of the vanilla is a long, elliptic capsule that takes up to 10 months to mature and splits in several places lengthwise as it ripens. The numerous round black seeds are contained within a sticky pulp.

The seed can be collected along with the capsule once is has matured. The sweet vanilla aroma is an instant giveaway that the fruit is ripe.

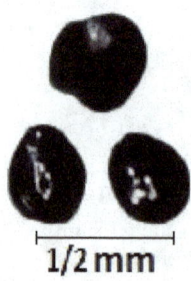

1/2 mm

Vanilla Orchid
Vanilla planifolia

Family: PAPAVERACEAE
Common name: **Poppy**
Number of genera: 44
Number of species: 760
Origin: Worldwide
Plants: Herbs, a few trees, and shrubs

The fruit is a capsule containing numerous small seeds, which are often light grey or brown to black in color, and sometimes pitted. The seeds can be collected as the capsule dries and begins to split.

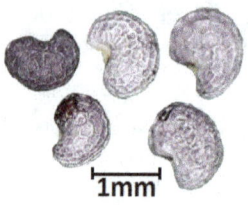

1mm

Poppy
Papaver somniferum

Family: PAPILIONACEAE
Common name: **Legume, Pea,** or **Pulse**
Number of genera: 400
Number of species: 12000
Origin: Worldwide
Plants: Trees, shrubs, herbs, and vines

The legume family is formally known as FABOIDEAE, FABACEAE, or LEGUMINOSAE.

The fruit is a pod called a legume with two equal halves that open upon drying, sometimes dispersing the seed with intense force.

The seed size and shape is variable, generally with a round or a flattened oval design, and sometimes with hairs.

Collect the pods individually or in clusters and allow time to dry before threshing. Some pods require rough treatment to release their seed. Wetting the pods sometimes helps to free the seeds.

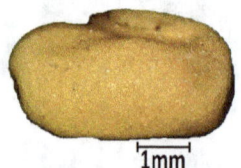

1mm

Fenugreek
Trigonella foenumgraecum

Family: PEDALIACEAE
Common name: **Sesame**
Number of genera: 13
Number of species: 50
Origin: Tropical to subtropical
Plants: Small shrubs and herbs

The fruits are non-fleshy capsules containing numerous seeds or a nut. The seed can be collected as the capsule dries. Sometimes threshing is required to release the seed. Sieve the chaff from the seed before storing.

Sesame *Sesamum indicum*

Family: PIPERACEAE
Common name: **Pepper**
Number of genera: 8
Number of species: 3600
Origin: Tropical to subtropical
Plants: Small trees, shrubs or scrambling vines

The fruits are fleshy single seeded drupes. The entire fruit is collected and dried; the seed is not usually removed from the fruit.

Black Pepper *Piper nigrum*

Family: PLANTAGINACEAE
Common name: **Plantain**
Number of genera: 3
Number of species: 270
Origin: Worldwide
Plants: Herbs with some shrubs

The fruits are fleshy to non-fleshy capsules or nuts. The seeds are collected as the capsules dry. Sometimes threshing is required to release the seed. Sieve the chaff from the seed before storing.

Plaintain *Plantago major*

Family: POACEAE
Common name: **Grass**
Number of genera: 760
Number of species: 10,000
Origin: Worldwide
Plants: Grasses

These plants have immense economic importance due to the number of agricultural crops within the family. The fruits of grass are classified as grains; they are different from other fruit because the seed coat is part of the fruiting body wall.

The seed can be threshed out of the chaff by manual and mechanical means. The seed is variable depending on the genus and can be round, oval, cylindrical, or have striations, pitting, or bristle-like projections.

19

Lemon Grass
Cymbopogon citratus

Family: POLYGONACEAE
Common name: **Knotweed**
Number of genera: 30
Number of species: 1000
Origin: Temperate Northern
Hemisphere.
Plants: Herbs, shrubs, and vines

The fruit is a nut which is brown to black in color. Light threshing of these seeds is sometimes required, and only a minor amount of sieving is needed to remove any immature seed and some chaff.

Sorrel *Rumex acetosa*

Family: PORTULACACEAE
Common name: **Pigweed**
Number of genera: 20
Number of species: 500
Origin: Worldwide
Plants: Annual or perennial herbs

The fruit is a capsule with one to numerous smooth or pitted black seeds. Collect part or all of the plant and spread on a ground sheet to dry. Alternatively, place the ground sheet

under the plant and collect the seeds as the capsules open.

1mm

Purslane *Portulaca oleracea*

Family: PRIMULACEAE
Common name: **Primrose**
Number of genera: 30
Number of species: 1000
Origin: Temperate Northern
Hemisphere
Plants: Annual or perennial herbs

The fruit is a small oval capsule opening at the top with minute three-sided brown seeds. Collect the capsules as they dry, and thresh to release the seeds. Sieve out the chaff and store the cleaned seeds.

Cowslip *Primula veris*

Family: ROSACEAE
Common name: **Rose**
Number of genera: 122
Number of species: 3350
Origin: Worldwide
Plants: Trees, shrubs, with some herbs

The fruits of the roses are extremely varied and can be pomes, drupes, hips, achenes, or other named fruit types. The seeds are also variable in size and shape, ranging from a single large seed to minute and dust-like small seeds.

Collect the mature fruits and remove the seed, depending on the fruit type. Other than the berry fruits, the seed of the drupes and pomes germinate best in the bottom drawer of a refrigerator wrapped in a moist tissue.

Coffee *Coffea arabica*

Family: RUTACEAE
Common name: **Rue**
Number of genera: 150
Number of species: 1500
Origin: Tropical to temperate
Plants: Evergreen trees or shrubs

The fruit is a berry or dry fruit that splits into parts. The seeds are compressed, oval, kidney-shaped, or elliptical, and are brown to black in color. Collect the fruits as they ripen and clean off any fleshy materials before drying. Sieve out any remaining chaff once the seeds are fully dry.

Salad Burnet
Poterium sanguisorba

Family: RUBIACEAE
Common name: **Madder**
Number of genera: 650
Number of species: 5000
Origin: Tropical to subtropical
Plants: Trees, shrubs, herbs, and vines

The fruit can be a dry capsule or a fleshy berry or drupe having two to numerous seeds which are often embedded in a succulent mass. The seed can be highly variable in size, shape, and coloration.

In this family, it is coffee that is of interest. Collect the coffee beans individually and remove the flesh before drying.

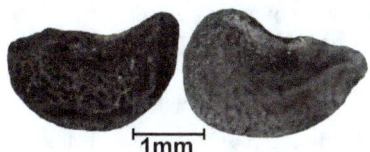

Rue *Ruta graveolens*

Lemon *Citrus limon*

Family: SAMBUCACEAE
Common name: **Sambucus**
Number of genera: 1
Number of species: 40
Origin: Tropical to temperate
Plants: Small trees, shrubs with some herbs

The fruit is a fleshy colorful drupe containing one seed. The collected drupes can be dried whole. Alternatively, the drupes can be placed in a sieve with slow running water and the flesh pressed gently through the sieve.

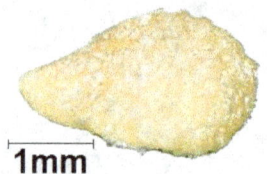

1mm

Black Elder *Sambucus nigra*

Family: SCROPHULARIACEAE
Common name: **Figwort**
Number of genera: 250
Number of species: 5000
Origin: Worldwide
Plants: Herbs and shrubs

The fruits are small capsules that split from the top or sides, producing few to many minute seeds. The seeds are arrow-shaped, kidney-shaped, rectangular, or oval; all are pitted and are brown to black in color. Collect the capsules and thresh to release the seed.

1/2mm

Mullein *Verbascum thapus*

Family: SOLANACEAE
Common name: **Nightshade**
Number of genera: 90
Number of species: 2800
Origin: Native to the Americas
Plants: Small trees, shrubs, herbs, and vines

Many genera within this family are important agricultural plants. The colorful fruits are fleshy capsules or berries which can be harvested upon maturity. The seed is flattened, curved, and oval, often with pitting. The color ranges from yellow, orange, grey, and brown to black.
Collect the fruits once fully ripe. The capsules may or may not open upon maturity. Dry capsules can be threshed, whereas fleshy capsules can be cut open and the seed removed manually.
The pulp of berries can be pressed gently through a sieve; alternatively, the seed can be placed in a container of water for a week or so to ferment, before being dried, then sieved clean.

1/2mm

Henbane *Hyoscyamus niger*

Family: THEACEAE
Common name: **Tea**
Number of genera: 40
Number of species: 600
Origin: Tropical
Plants: Large trees and shrubs

The fruit is a berry, drupe or capsule. Collect the fruit individually once mature and remove any flesh or pulp before drying and storage.

Green Tea *Camellia sinensis*

Family: VALERIANACEAE
Common name: **Valerian**
Number of genera: 13
Number of species: 350
Origin: Tropical to temperate
Plants: Herbs with some shrubs

The fruit is a group of closely-packed singular seeds known as an achene. Collect entire flower heads after the flowers have died. Store the heads in a paper or material bag and allow a week or two for the seed to dry fully prior to threshing and cleaning.

Valerian *Valerian officinalis*

Family: ZINGIBERACEAE
Common name: **Ginger**
Number of genera: 40
Number of species: 1000
Origin: Tropical and subtropical
Plants: All are herbs

The fruit is a capsule or fleshy seed-filled colorful berry. The seed could best be described as segments of a sphere; the color is brown to black. Collect the fruit and clean as per the fruit type encountered.

Cardamon
Elettaria cardamomum

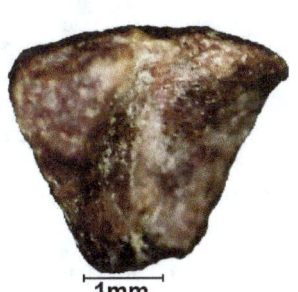

Native Ginger
Alpinia caerulea

CHAPTER 3
HOW TO THRESH AND CLEAN SEED

Threshing

First, let's look at the term **threshing**.

Threshing is the term used for the breaking up and separation of the dried seed pods, seed heads, or capsules from the seed. The waste from the seed is called **chaff**. It is after threshing that the actual cleaning of seed occurs. These are two separate activities. Threshing of one form or another is usually required prior to cleaning.

With the use of some common household items, the threshing of seed is not very difficult. Sieves, colanders, tea strainers, buckets, and bowls can all be employed to get the job done—even by professional collectors. Additional equipment for enthusiastic collectors includes a mechanical thresher or a garden blower-vac.

Threshing can be undertaken by a number of means, including manual and mechanical methods. Both are effective, although an appropriate method should be sought depending on the amount of material that requires threshing. There is little point in using a mechanical thresher for a handful of material; conversely, you really don't want to thresh a truckload of material by hand!

Hand Threshing

As implied by the name, this is the use of your hands to crush the chaff from around the seed. This can be undertaken by placing the material in a bucket or bowl, and crunching it up until the seed is freed from the capsules. Gloves are generally required for this method, as the action required can hurt your hands.

Unfortunately, this is not always an effective technique for many of the harder woody species, as the pods are simply too difficult to crush manually.

After you have finished hand threshing the seed / chaff material, gently but firmly swirl the contents around in a bucket or bowl. The usually heavier seed will work its way to the bottom, allowing the bulk of the chaff to be removed and discarded. This makes the cleaning process easier. Should the seed and chaff not separate; cleaning will be more complicated and time-consuming.

Threshing with Sieves

Threshing can be undertaken by simply pressing the collected material through an appropriate sieve. This works well for many species with relatively brittle capsules or pods.

To determine which sieve is best suited for the material being threshed, select a sieve with holes slightly larger than the seed.

Good quality gloves are recommended when using this method, as injuries can occur from small sticks, thorns, etc. piercing the skin.

Attaching a sieve to a child's swing can aid greatly in this method of threshing, as it enables you to use a much larger sieve.

Manual, hand operated thresher

Threshing with a Towel or Rag

This method is restricted for the seed of fruit and berry species where the outer layer of the seed peels away like a skin. The cucurbits are one such group where this method is well-suited.

To use this method, rub the seeds between two layers of material, such as calico or rags. Place only small amounts of seed between the materials at a time, and rub back and forth to remove the chaff.

After you are satisfied the seed and chaff are separated, pour everything into a container to be cleaned.

Mechanical Threshing

There are many mechanical threshers on the market, both manual and motorized. Regardless of type, they are expensive to purchase. You would need to be doing large quantities of threshing to make the purchase of one worthwhile.

Cleaning Seed

The cleaning of seed after it has been threshed can be a very rewarding experience, as you get to see the end product of your labors. Cleaning, at its most basic, needs only to involve the use of two bowls and the wind; at the more complicated end of the spectrum, it may employ specialized equipment, such as sieves or a motorized cleaner.

Regardless of what equipment is required, cleaning the seed helps reduce the space required for storage, reduce or eliminate pest problems, and makes replanting much easier.

Two Bowl Method

The most simple and inexpensive way to clean seed is the aptly named Two Bowl Method. First, find two suitable bowls; these can be medium-sized plastic food containers, 10-liter water buckets, etc. Use your imagination!

Slowly tip the uncleaned seed (seed and chaff mixture) from the first bowl into the bowl underneath, allowing the wind to blow the chaff away. Wind can be provided either naturally or via a fan. The distance between the bowls will require adjustment, depending on wind strength. Swap bowls and continue tipping the seed from one bowl to the other until you are satisfied with the cleanliness of the seed. In some cases, the seed will not become 100% clean, and a final clean using tweezers is required. Amaranth and eucalyptus are good examples of seeds that are left only partly clean using this method.

Flour sieves

Sieves and Sieve Sizes

Having several sieve sizes is always helpful, although purchasing an analytical quality sieve (as pictured) can be prohibitively expensive.

Assorted kitchen sieves

Analytical sieve

There are, however, alternatives to using expensive analytical sieves. The most cost-effective sieves can be found in almost every kitchen. These are basic flour and kitchen sieves, tea strainers, and colanders, all of which are available at your local store.

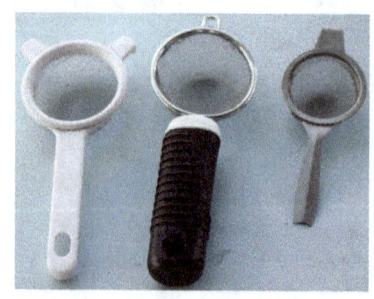

Assorted tea strainers

Other Sieves

There are many household items that can be used effectively as sieves for cleaning seed. Mosquito meshing comes in a handy size, whilst a colander is excellent for some of the larger seeds. Audio speakers, both car and home, provide mesh that makes an excellent sieve.

The material used for the mesh of a sieve is limited only by your imagination and creativity.

Speaker cover

Cleaning Pods and Dry Capsules

After capsules or pods have been threshed, the cleaning is greatly expedited using sieves. Often two or more sieves are used to separate the larger and smaller portions of the chaff from the seed.

Use several wide-mouthed bowls whilst sieving the seed, and have a bucket handy into which to place the waste. Doing this helps in chaff separation, and also contains any spillage

in the event of an accident, making re-sieving easier.

Start with a sieve that has a mesh size several sizes larger than the seed. The reasoning behind this is to remove the larger unwanted materials as efficiently as possible; this also reduces the volume of materials remaining to be cleaned.

After the bulk has been removed, use a sieve that is only slightly larger than the seed and slowly sieve the seed through. There will usually be several seeds remaining that are bigger than usual; collect these to add back to the cleaned seed once finished. Repeat this process until no large chaff can be effectively removed.

Now use a sieve slightly smaller than the seed and sieve out the smaller material. Often, small seed will be sieved out. Reject these, as they are usually immature or inferior. It is more beneficial to focus on the larger, healthy seed.

Repeat this process until the seed is as clean as possible. The final clean can be undertaken utilizing the wind or a mechanical cleaner.

Cleaning Fruits and Berries

The cleaning of fleshy fruits and berries is sometimes more interesting than it sounds, as many seed-bearing fruits have their own unique methods of seed extraction.

Berries

Berries are fruits with more than one seed, such as henbane, caper and pomegranate. Some berries only have a few seeds, while others will contain hundreds. Regardless of how many seeds a berry has, they can usually be cleaned using a sieve and water.

If the size of the seed is unknown, you will need to gently cut open the berry and check the seed. To determine what size sieve will be required, select one that is slightly smaller than the seed. This will allow the pulp of the berry to be washed through the mesh. You may need to press the pulp through the sieve with your fingers or a small spatula.

Once the sieve is selected, place the entire berry or sections of it on the sieve. This process will depend on the size of the berry and sieve. Do not overload the sieve, as it only makes the task more difficult, and may result in the loss of seed, the failure of the sieve, or both.

With the water running slowly over the berry, gently press the pulp through the sieve. The seed should remain in the sieve. If the seed also passes through the sieve, select a smaller mesh size and start again.

Once the seed is considered clean, and the pulp is removed, place the seed on a clean, non-plastic surface and allow it to dry. This process can be expedited by dabbing the seed with a dry cloth, which can be a little tricky if the seed is small, as it tends to stick to the cloth and become a problem.

Spread the seed thinly and avoid mounds, as these are more likely to become moldy. As the seed dries, stir or mix at least daily to aid in thorough drying. This method works well for both small and large seeded species.

Drupes

Drupes are single-seeded fruits such as neem, olive, and nutmeg. These can be either fleshy or not; this characteristic will determine how the seed is cleaned.

A quite pleasurable way to clean some of the fleshy edible species is to eat the flesh. The fleshy drupes can be cleaned with a sieve in the same manner as berries.

However, cleaning the drupes without flesh can sometimes be much more difficult.

The very small drupes with dry flesh can be cleaned whole. The entire fruit is left to dry out thoroughly prior to threshing the skins off and then cleaning with a cleaner or sieves.

CHAPTER 4
THE STORAGE OF SEED

Introduction

Storing seed correctly is the most important step when considering how to collect, clean, and store seed. If seeds are stored correctly, they can last for years. However, if stored poorly, they will lose their viability and fail to germinate once planted.

An important point to remember here is that there are a number of plant species that will only remain viable for a short time, i.e. *Syzygium* sp., which will only remain viable for a few weeks. Long term storage of these seed is not possible so they should be planted as soon as possible.

There has been a great deal of research conducted over the last two centuries into the viability of seed under storage conditions, and many good research papers have been published on the findings. This research has shown that many of the seed species that are collected and cleaned correctly can be stored for use at a later date. This is good news for gardening enthusiasts, as it means that most of the plants they are collecting seed from can be preserved.

Regardless of the challenges involved in the storage of seed, it can be a very rewarding experience. All keen gardening enthusiasts should attempt to store the seed from their respective interests to plant in later seasons or exchange with other like-minded enthusiasts.

Take notes on when the seed was stored, how it was stored (in a box, in the refrigerator, etc.), and most importantly, when it was planted and if

it germinated. This information on the viability of plant species in your district is very important to plant researchers, as it contributes to the general knowledge on plants worldwide.

Putting Seed into Storage

Only store the best quality seed; i.e., seed that is undamaged, free of defects, pests, chaff, and debris. There are occasions when you cannot collect the best specimens, as the seed simply isn't available due to weather conditions, seasonal factors, or just bad timing. When this occurs, all you can do is the best you can with what you have. Sometimes the seed will be suitable to propagate next season, and sometimes it will fail. This is all part of the enjoyment of collecting seeds.

Seeds should be stored once they are cleaned of all chaff, dirt, and fleshy dried pieces. This helps reduce the volume being stored and reduces the amount of contaminating materials in the seed. Contamination includes both pests and loose materials. Pests include insects, molds, and fungi, all of which can easily destroy stored seed.

Seed should not be put into storage on days when the humidity is high, as the seed may draw moisture from the atmosphere, which can lead to spoilage. This most commonly occurs when seed is stored in plastic bags.

Ideally, seed should be stored once fully dry. This sounds simple enough, but looks can be deceptive, and many seeds have been ruined due to eagerness. Most seeds take about two weeks to fully dry if maintained at an even, warm temperature. They may look dry on the outside, but are often still moist on the inside, so be patient. Many seeds reduce in size as they dry, so also look for this as it occurs.

Once the seeds are ready to be stored, take the time to examine them carefully for any problems, and remove anything that looks suspicious. "When in doubt, throw it out" is a simple expression that is well worth remembering.

Overcoming Seed Pest Problems

The storage of seed has always been an issue due to the many and varied circumstances that can arise without warning. Issues such as humidity, hot summers, cold nights, insects, molds and fungi, and many others add to the difficulty in storing seed successfully.

Seed for storage must be dry, free of dirt, and most importantly, free of insects and their eggs. The weevil eggs pictured below have hatched and the larvae have eaten this bean seed.

Weevils and Other Beetles

An excellent way of removing insect eggs from seed, especially large hard seed such as beans, is to make up a dilute solution of bleach (10ml/L) and soak the seed for ½ an hour. Remove and rinse the seed before thoroughly drying.

Fungicides and insecticides are readily available and can be used to treat seed where problems are too

intense to overcome without their use. Take every possible precaution when using these products, as they are **dangerous**. Read the instructions on the label before using any product.

Weevils are one of several insect pests that eat seed. The photo above shows the damage that can happen if seed is stored poorly. A mature weevil is shown below.

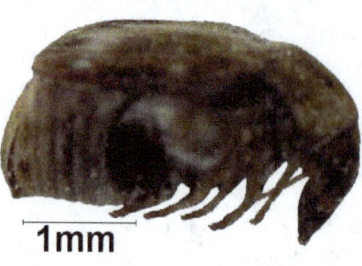

Weevils and other seed-boring beetles can range in size from almost microscopic to large, impressive specimens. Wood- and seed-boring beetles range in color from drab brown to highly colorful, and many are highly prized by collectors.

Moths

Moth larvae are a major economic pest in cereal crops throughout the world. A great deal of money is spent each year in trying to control the damage and losses caused by these small, troublesome insects.

The specimen shown below is about to pupate after consuming the bulk of a Bauhinia seed. Also shown is an immature larva that was inside the nearby seed when the pod was opened.

Being observant when cleaning your seed will prevent many of the problems that can arise with seed storage. However, it is always a good practice to check your seed regularly and remove anything that may look suspicious.

Fungi

Fungi are the group of plants that lack chlorophyll and leaves. They include molds, mushrooms, mildew, rusts, and smuts.

Of the pest problems encountered, problems with fungi are often the easiest to control and manage.

The problem with molds usually occurs if the seed is put into storage while it is still moist, or if the seed is stored under poor environmental conditions. Whatever the case, if the situation is discovered soon enough, the problem may be reversed, as it

takes some time for molds and fungi to penetrate the seed surface and cause damage.

If seed is found with mold growth, you should first remove all affected seed lots from the storage container, as you do not want to infect the remainder of your collection. The affected seed can either be disposed of or cleaned. If the seed is not too badly contaminated, the problem can be solved without disposal.

To clean mold from seed, gently rub the seed with a cloth moistened with bleach solution (10ml/L dilution) until the mold is removed. Staining often remains, but this should not affect the seeds' ability to germinate. This problem will have affected the seeds' long-term storage viability, so replacement should be considered as soon as possible.

Finally, fungicidal powders can be purchased that will inhibit fungal growth. These should be considered if this problem is likely to occur in your location. Remember to read the label and always follow the manufacturer's instructions—these products are **dangerous** when used incorrectly.

Temperature

Few people have access to expensive storage equipment that can maintain a stable temperature at the desired temperature setting. There are two issues here: the temperature at which the seed is stored, and the stability of this temperature. Seeds can be maintained at a low temperature of 2-4^{0}C for long periods, or they can be frozen for storage over centuries. It is the stability of the temperature that is of the greatest importance to the average collector.

As there is no way of maintaining a constant temperature without equipment, we will look at how to maintain a stable temperature using some commonly found items.

Changing temperatures are not good for seed storage. The constant fluctuations from the day and night cycle will cause the most rapid decline of seed viability. This is one problem that must be overcome if you intend storing seed for any length of time.

To achieve optimal results in saving seed, a suitable position with a stable temperature must first be selected. Avoid locations such as external walls and areas near amenities, air conditioners, fireplaces, or any other temperature-altering devices.

Once a storage location is selected, a suitable container in which to house the various lots of seed must be obtained.

Storage containers

Containers that are suitable include, but are not limited to: foam boxes, cardboard boxes, buckets, fiberglass containers, drums, etc. Some storage containers will require modifications to overcome temperature and humidity changes.

Foam boxes are the best choice, as they are already insulated and are often fully sealed. They can be found in numerous sizes, and usually have tight-fitting lids.

Cardboard boxes are good choices; however, finding thick boxes without gaps at the bottom and with tight-fitting lids is sometimes difficult. Additional cardboard can be glued in any places with gaps and can be used to thicken the base, top, and sides as required. Alternatively, line a cardboard box with polystyrene foam—thereby providing the best of both worlds.

Buckets and drums with tight-fitting lids can be used to store seed as long as the inside is properly insulated with foam or other suitable material. Remember that both the bottom and top require insulation. Otherwise, changes in temperature and humidity will still occur inside the container.

Polystyrene and cardboard are excellent insulation materials. Be extremely careful in choosing other insulating materials, such as roof and wall insulation. These items are not designed to be disturbed and are dangerous if handled incorrectly. If you require insulation materials for your chosen container that may be harmful, you should consult an industry expert.

Humidity

As changes in humidity can cause the loss of seed viability, the container that you have chosen to maintain a stable temperature must also be capable of preventing changes in humidity. This is one reason that the lid must be tight-fitting.

Humidity within any dwelling can change dramatically during the daily cycle, especially in tropical regions during the wet season. Even small changes in humidity can have unwanted results on the viability of your stored seed.

Silica gel is a desiccant, and is excellent for protecting your seed from moisture. Some gels are coated with cobalt chloride or other indicators that reveal when their usefulness has expired. You should check these regularly.

All silica gels can be dried and reused numerous times. To do this, place the gel packs on a tray and place in a cooling oven overnight. Ensure the oven is not too hot, as you do not want to burn the packaging of the gel packs.

Silica gels in one form or another can be purchased from chemists, hardware stores, nurseries, and supermarkets. There are a wide range of products on the market, and listing them all would be almost impossible. Water storage granules from nurseries are ideal.

Silica gel packs can also be collected from tablet bottles and within the packaging of electrical items and other miscellaneous products.

Remember that not all seeds can be stored, so choose your seeds carefully and avoid those that will present problems. As your experience grows, you can become more adventurous and attempt to store more difficult species, taking notes about the methods used and the outcomes.

Many of the rare species in collections around the world today have only survived because plant enthusiasts have saved them. Your efforts could lead to the saving of species for future generations to appreciate.

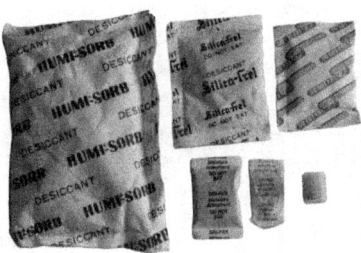

Gel packs

Conclusion

The storage of seed can sometimes be challenging. However, the effort is very rewarding, and many varieties of seed can be stored successfully by anyone willing to take the time and effort.

Following simple suggestions, being patient, and taking care to avoid contamination and pests will all lead to the successful storage of seed.

REFERENCES

Auld B.A. & Medd R.W., *Weeds. An illustrated botanical guide to the weeds of Australia.* Inkata press, 1992

Bodkin F., *Encyclopaedia Botanica.* Cornstalk publishing, 1992

Chevallier A., *The Encyclopedia of Medicinal Plants.* Dorling Kindersley Limited. 1996

Culpeper's Colour Hedrbal. W. Foulsham & Company Limited, 1983

Fanton M & J., *The Seeds Savers Hndbook.* The Seed Savers' network, 1993

Herbs, The Essential Guide for a Modern World. Rodale International Ltd, 2006

Hoffmann D., *Holistic Herbal.* Element Books Limited, 1998

Hutchinson C. *The Australian Home Orchard Growing Fruit in Your own Garden.* Simon & Schuster Australia, 1993

Newdick J., *Book of Herbs.* CLB Publishing,m 1994

Norman J., *The Complete Book of Spices. A practical Guide to Spices & Aramatic Seeds.* Dorling Kindersley Limited, 1991

Ortiz E.L., *The Encyclopedia of Herbs, Spices & Flavourings.* Carroll & Brown Ltd, 1993

Pilcher M, Davis L, Hurrion D., *Garden Terms.* Hamlyn, 1995

Magic and Medicine of Plants. Readers Digest, 1994

Herbs, The essential 21st Century Guide. Rodale, 2006

Sanecki K. N., *The Book of Herbs.* Quintet Publishing Limited, 1985

Shipard I., *How can I use Herbs in my Daily Life?,* Queensland Complete Printing Services, 2003

Shipard I., *How can I grow and use Sprouts as living food?,* Queensland Complete Printing Services, 2005

Shipard I., *How can I be Prepared with Self-Sufficiency and Survival Foods?,* Queensland Complete Printing Services, 2008

Wikipedia, the free encyclopedia

INDEX

A
ACANTHACEAE,
Achene, 4
Achillea millefolium, 9
Allium schoenoprasum, 5, 15
Allspice, 17
Aloe vera, 15
Alpinia caerulea, 23
Althea officinalis, 16
ANACARDIACEAE, 8
Anethum graveolens, 6, 8
Angelica, 9
Anise, 9
Anthriscus cerefolium, 8
APIACEAE, 8
Apium graveolens, 9
Archangelica archangelica, 9
Artemisia vulgaris, 10
ASTERACEAE, 4, 9

B
Barbados Aloa, 15
Barbarea vulgaris, 10
Bean, 11
Basil, 13
Berry, 4
Black Elder, 22
Black Pepper, 19
Borage, 10
BORAGINACEAE, 10
Borago officinalis, 10
BRASSICACEAE, 6, 10
Brassica juncea, 6, 10
Brown Mustard, 10

C
CAESALPINIACEAE, 11
Callitris intratropica, 12
Camellia sinensis, 23
CANNABACEAE, 11
Cannabis, 11
Cannabis sativa, 11
Caper, 11
CAPPARACEAE, 11
Capparis spinosa, 11
Capsella bursa-pastoria, 10
Capsule, 4
Caraway, 9
Cardamon, 23
Carum carvi, 9
CARYOPHYLLACEAE, 11
Cashew, 8
Caster Oil Plant, 12
Catmint, 14
Cedar, 12

Celeriac, 9
Cereal Crops, 5
Chervil, 8
Chickweed, 12
Chives, 5, 15
Cichorium intybus, 9
Citrus limon, 5, 21
Cleaning seed, 25
   Berries, 28
   Drupes, 28
   Dry capsules, 27
   Fleshy fruits, 27
   Sieves, 26
   Two bowl method, 25
Coffea Arabica, 4, 21
Coffee, 4, 21
COMPOSITAE, 9
Coriander, 9
Coriandrum sativum, 9
Cowslip, 20
Cress, 10
CRUCIFERAE, 10
Cryptotaenia candensis, 9
Cumin, 9
Cuminum cyminum, 9
CUPRESSACEAE, 12
Cymbopogon citrates, 20
Cypress Pine, 12

D
Daisy, 4, 9
Dill, 6, 8
Drupe, 5
Drupelet, 5

E
Echinacea, 10
Echinacea purpurea, 10
Elettaria cardamomum, 23
Epilobium billardierianum, 17
Etaerio, 5
EUPHORBIACEAE, 12
Evening Primrose, 17

F
FABACEAE, 18
FABOIDEAE, 18
Fennel, 9
Fenugreek, 18
Figwort, 22
Flannel Flowers, 8
Flax, 15
Foeniculum vulgare, 9
Fragraria vesca, 5

G
GERANIACEAE, 13
Geranium, 13
Geranium robertianum, 13
Germen Chamomile, 10
Ginger, 23

Grain, 5
Grass, 19
Green Tea, 23

H
Hemp, 11
Henbane, 22
Henna, 15
Herb Robert, 13
Hesperidium, 5
Hip, 5
Horehound, 14
Horseradish Tree, 16
Hyoscyamus niger, 22
Hyptis suaveolens, 14
Hyssop, 13
Hyssopus officinalis, 13

I
Iboza riparis, 14
ILLICIACEAE, 13
Illicium verum, 13

J
Java Tea, 13

K
Knotweed, 20

L
LABIATAE, 13
LAMIACEAE, 13
Lateriflora scutellaria, 14
Lavender, 14
Lavendula vera, 14
Lawsonia inermis, 15
Legume, 6, 18
LEGUMINOSAE, 18
Lemon, 5, 21
Lemon Balm, 13
Lemon Grass, 20
Leonurus cardiaca, 13
Levisticum officinale, 9
LILIACEAE, 14
Lily, 14
LINACEAE, 15
Linseed, 15
Linum usitatissimum, 15
Loosestrife, 15
Lovage, 9
Luffa, 6
Luffa acutangula, 6
LYTHRACEAE, 15

M
Madder, 21
Mallow, 15
MALVACEAE, 15
Mahogany, 16
Margoram, 14
Manihot dulcis, 12
Marshmallow, 16

35